D1297468

About the Author

RUSSELL LYNES (1910–1991) was an art historian, cultural critic, author, photographer, and managing editor of *Harper's Magazine*. His articles for *Harper's* and *Life* in 1949 made parsing American culture into highbrow, middlebrow, and lowbrow a national pastime. He wrote many books, including *Guests* and *The Tastemakers*.

Snobs

Also by Russell Lynes

Guests

The Tastemakers

A Surfeit of Honey

Cadwallader

The Domesticated Americans

Confessions of a Dilettante

The Art-Makers of Nineteenth Century America

Good Old Modern

More Than Meets the Eye

The Lively Audience

Life in the Slow Lane

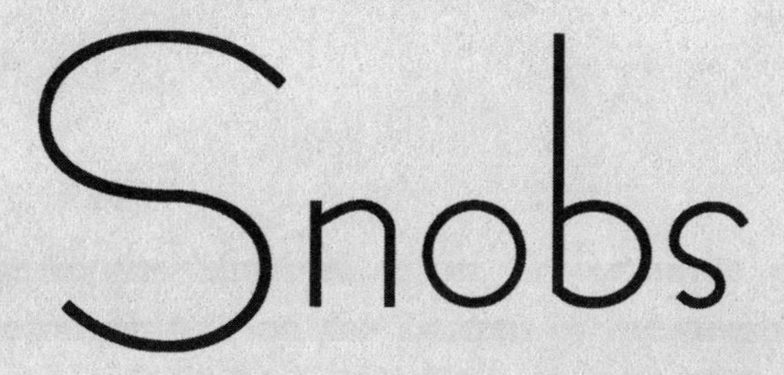

The Classic Guidebook to Your Friends,
Your Enemies, Your Colleagues,
and Yourself

RUSSELL LYNES

With Drawings by Robert Osborn

HARPER

NEW YORK • LONDON • TORONTO • SYDNEY

HARPER

A hardcover edition of this book was published in 1950 by Harper & Brothers.

FIRST HARPER PAPERBACK PUBLISHED 2009.

Library of Congress Cataloging-in-Publication Data is available upon request.

ISBN 978-0-06-170640-0 (Harper paperback)

09 10 11 12 13 /CW 10 9 8 7 6 5 4 3 2 1

For

M. A. L.

INTRODUCTION

Joseph Epstein

The mechanics of snobbery are simple enough: a snob seeks ways to establish his superiority over his fellow human beings. He sees it as one of his main tasks in life to make it known that he is, for reasons he should be only too pleased to supply, better than you or me. Snobs come in two main kinds: downward-looking, who feel themselves well established and enjoy looking down on others, and upward-looking, who yearn to elevate their position in life so that they also might look down on others. Snobs of both kinds can be amusing to watch, at least from the middle distance, but are less than pleasing to be with. As the Jews of Russia used to say of the Tsar, so one tends to think of a snob: he should live and be well, but not too close to me.

Snobbery, being part of human nature, does not change. Snobberies, however, are subject to fashion and therefore change frequently. The largest change in snobbery over the past fifty years in America—which means that it occurred since Russell Lynes wrote his book—is that having to do with lineage, or ancestry. Until the late 1960s, with its head-on attack on all things establishmentarian, much snobbery flowed from what used to be thought of as WASP culture. WASP stands for White Anglo-Saxon Protestant, but that isn't the tenth of it. The true WASP was Eastern seaboard, possibly at the outside a Virginian, who attended a small number of okay Ivy League schools, banked and invested at certain exclusive institutions, used a small number of white-shoe law firms, and tended to marry strictly his or her own kind. The WASP was the American version of the English aristocrat, and like his British counterpart, he or she was the figure from whom all major snobberies flowed.

Once the primacy of the American WASP was over, snobbery itself went, of all things, democratic, which means that we are all snobs now. Suddenly Catholics, Jews, African-Americans, and other ethnic groups, all of whom previously felt they didn't quite have a seat at the main table in American life, came to take a deep pride—some might say an almost snobbish pride—in their difference from what was once thought to be mainstream culture. So completely had the tables turned that suddenly the WASPs were the only group in America about whom everyone was free to take critical shots.

Russell Lynes is not a grand theorist of snobbery—he does not, for example, write at any length of the psychology of the snob—but its American taxonomist. His book is devoted to delineating the major categories of snobbery and then describing the multiple forms that snobbery has taken in American life, including what he calls "the Anti-Snob Snob," also known as the Reverse Snob, or the person "who finds snobbery so distasteful that he or she is extremely snobbish about nearly everybody since nearly everybody is a snob about something."

Almost every instance of snobbery that Lynes described more than half a century ago is still in business today. But then, as every subject has its politics, so does every subject have its potential snobbery. Behind much contemporary snobbery is an element of self-congratulation. This is most notable when politics and snobbery combine in the character known as the virtucrat: the person whose snobbery, a variant of Lynes's category of the Moral Snob, derives from his certainty that his virtue is greater than yours because, or so he believes, he cares more about social injustice or global warming or fighting AIDS than you or, really, just about anyone else.

Russell Lynes chose not to be a Snobographer, or biographer of snobs, a subject on which vast tomes could be written. Few are the proper names mentioned in *Snobs*, and fewer still the anecdotes told about snobs. My own favorite anecdote in this line has to do with the publisher Alfred A. Knopf, who was an earnest wine snob. When Alfred's brother invited him to dinner at his home, the brother took great care to select a costly bottle of wine. After properly decanting the wine at just the right temperature, allowing it to breath for precisely the right amount of time, he poured Alfred a glass and awaited his verdict. Since none was forthcoming, he asked, "The wine, Alfred, how do you find the wine?" "Oh," replied Alfred Knopf, "served in these glasses, how can I possibly tell?"

Russell Lynes ends his book by suggesting that, in one matter or another, we are all snobs, including those of us who have written books about snobbery, and he is, of course, right. Nothing for it but to admit to our snobberies upfront, at least to ourselves; do what we can to conquer them; and get on with life as graciously as we can.

Snobs

THERE was a time not long ago when a snob was a snob and as easy to recognize as a cock pheasant. In the days when Ward McAllister was the arbiter of Newport society and when there were precisely four hundred souls in New York worth knowing and only "nobodies" lived west of the Alleghenies, snobbishness was a nice clean-cut business that made careers for otherwise unoccupied women and gave purpose to otherwise barren lives. In those days the social order was stratified as tidily as the terracing of an Italian garden, and a man could take his snobs or leave them. But now the social snob, while not extinct, has gone underground (except for professionals such as headwaiters and metropolitan hotel room-clerks), and snobbery has emerged in a whole new set of guises, for it is as indigenous to man's nature as ambition and a great deal easier to exercise.

The Social Snob has gone underground (except for professionals such as headwaiters and roomclerks).

SNOBS

Snobbery has assumed so many guises, in fact, that it is, I believe, time that someone attempt to impose order on what is at best a confused situation. There are a few basic categories of snobs that seem to include most of the more common species that one is likely to encounter, or, indeed, to be. None of these categories is new; there have always been, I presume, snobs of many types,* but now that the preeminence of the social variety has been submerged in a wave of political and economic egalitarianism, and now that we find ourselves in an era in which the social scientists believe that it is somehow good for us to be ticketed and classified, let us sort out the most common practitioners of the sneer.

The Intellectual Snob is of such distinguished lineage and comes from such established precedent that he is dignified by a mention in Webster's ("one who repels the advances of those whom he regards as his inferiors, as, an intellectual snob"). The other categories are less well known and less well documented. For convenience, let us call them: the Regional Snobs, the Moral Snobs, the Sensual Snobs, the Emotional Snobs, the Taste Snobs, the Occupa-

* It is 102 years since William Makepeace Thackeray published his *Book of Snobs*, a series of facetious essays that originally appeared in *Punch*. Mr. Thackeray's snobs are largely of the social sort.

tional Snobs, the Political Snobs, and finally the Reverse Snobs or Anti-Snob Snobs. Before we examine these, we should be aware that economic and social boundaries, while they may occasionally serve as guide ropes, are on the whole unimportant in considering the various forms of condescension and the various attitudes of superiority that distinguish the true snob from the merely vain man, woman, or child.

Snobbishness, as we will use the word, implies both an upward and a downward movement—a scramble upward to emulate or outdo those whose position excels one's own, and a look downward on (or sometimes straight through) those less happily endowed than oneself. The true snob never rests; there is always a higher goal to attain, and there are, by the same token, always more and more people to look down upon. The snob is almost by definition insecure in his social (in the larger sense) relationships, and resorts to snobbishness as a means of massaging his ego. Since scarcely anyone is so secure that his ego does not sometimes need a certain amount of external manipulation, there is scarcely anyone who isn't a snob of some sort. As a matter of fact, the gods of the Greeks and the Romans were frightful snobs, morally, physically, and emotionally, and it is not uncommon for civilized peoples to worship snobbery. It is the Christian religion that promoted the virtue of humility for us, and of all the virtues it is the most difficult to come by. Let us not, then, be snobbish about snobs—at least not yet.

It is not my intention to apply the scientific method to the definition of the categories which we shall examine, though each species will be seen to have its subspecies and each subspecies to have many variants. The reader will discover in my approach certain similarities to that dark medieval yardstick, the Seven Deadly Sins, a once useful method of classifying man's shortcomings. He may also see the Four Temperaments or Humors as dominating some of the groups we look at. It will, I trust, become apparent that each snob suggests another snob, as each sin another sin, each temperament another temperament. It is not possible here either to exhaust the possibilities or even to give those we examine more than a casual nod. I mean this to be suggestive, merely a sketch that will make the reader glimpse the vast possibilities that a methodical study by a diligent social scientist might uncover.

Regional Snobs

Our first category is the Regional Snobs, commonly known in the South as Virginians, in the West as Californians, and in the East as Bostonians. This, how-

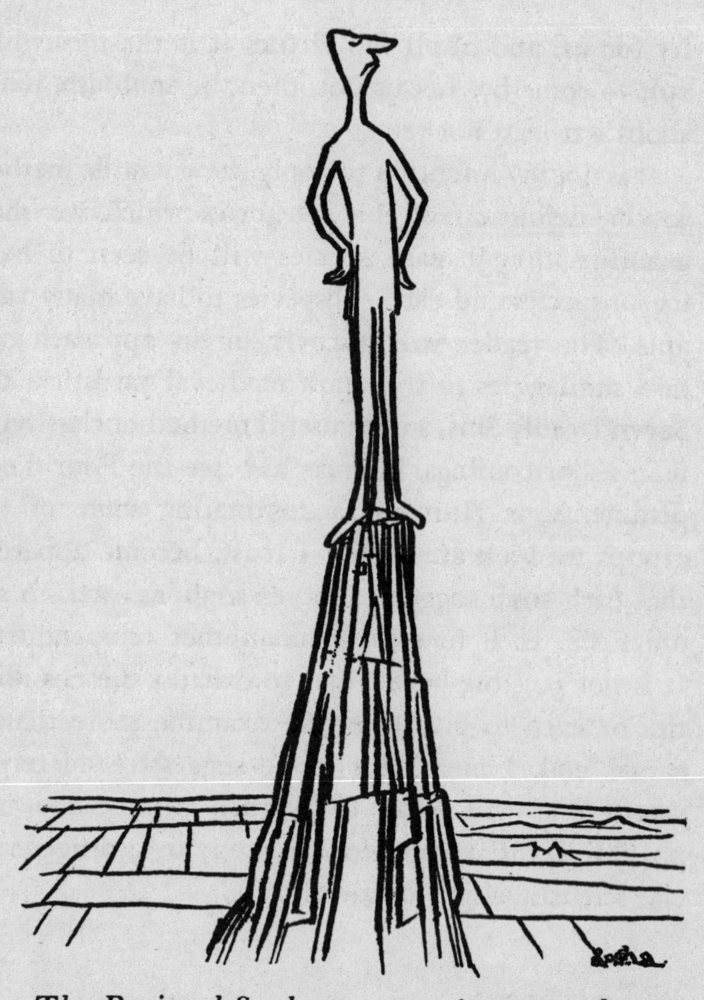

The Regional Snob can come from anywhere.

ever, should be recognized for what it is, a mere colloquialism. The Regional Snob can come from anywhere, and is readily distinguished by his patronizing attitude toward anywhere else. He lets it be known that there is no place to match the seat of his origin; indeed, he seems surprised or amused that people in other places are so much like people. The Asturians who live in the north of Spain, for example, look with special distaste on the citizens of the neighboring province, Galicia, and they have a saying that "a Galician is the animal that most closely resembles a human being." In Texas it is said that you should never ask a man where he comes from. "If he's a Texan," they say, "he'll tell you. If he's not, don't embarrass him." These are not as extreme cases as they might seem. It was recorded a decade ago that a boy who lived on Martha's Vineyard, an island off the Massachusetts coast, was assigned the problem in school of writing a composition about the then Duce of Italy. His paper started with the sentence: "Mussolini is an off-Islander."

But let us consider more common types of Regional Snobs. In Vermont, for example, the Regional Snob is generally called a "native" to distinguish him from the group known as "summer people." The aloofness of the Vermont native, a man proud of his thrift, of the bleakness of his winters, and especially of the fact that he has managed to squeeze a living out of rocky hillsides and out of "summer people,"* has a special laconic quality that is guaranteed to freeze the marrow of, say, a Texan. This kind of Regional Snobbism is of the *We've had it tougher than anybody* variety, and is the opposite of the California type which is of the *We know how to live better than you do* kind or of the Gracious Living types found in the South, notably in Virginia, in South Carolina, and in the New Orleans vicinity.

These types are, more or less, Area Snobs and should be distinguished from the Local or Home Town varieties which demonstrate certain cultural patterns quite different from those found in general geographical areas. The Local Snob does not even in many cases recognize his home town as anything very special; his vision may be myopic to the extent of permitting everything beyond the end of his particular street to go out of focus. "The other side of the tracks" is a phrase less frequently heard than it was a generation or so ago. We live in an age of "developments"—real estate developments, housing developments, community developments—of "projects" and

* and more recently, with the advent of the Ski Snobs, out of "winter people" as well.

of subdivisions, and the railroad tracks have lost some of their social significance in this age of busses and automobiles. So we have subdivision dwellers looking down upon development dwellers, and development dwellers turning their heads away from project dwellers, and project dwellers scornful of tenement dwellers. But the genuine Home Town Snob is rather more special than any of these.

Boston is too well known for its special brand of provincial hauteur to need discussion here, but the New York brand is less well documented and will serve to demonstrate one of the extreme forms of Local Snobbism. This is the Cultural Capital variety, or *Anything or anybody of any interest comes here* kind, that makes the New Yorker when visiting in any other city assume an air of condescension that has both an overhead spin and a reverse twist. "You know," the New Yorker* will say when visiting a city in the Middle West, "I think it's really terribly interesting *out here.*" It is a wonder that so few New Yorkers get their throats cut in what they think of as (but do not call) "the provinces." In its most advanced forms Cultural Capital Snobbism will bend

* Not to be confused with the magazine of the same name. It is not within the scope of this essay to discuss institutional snobbism.

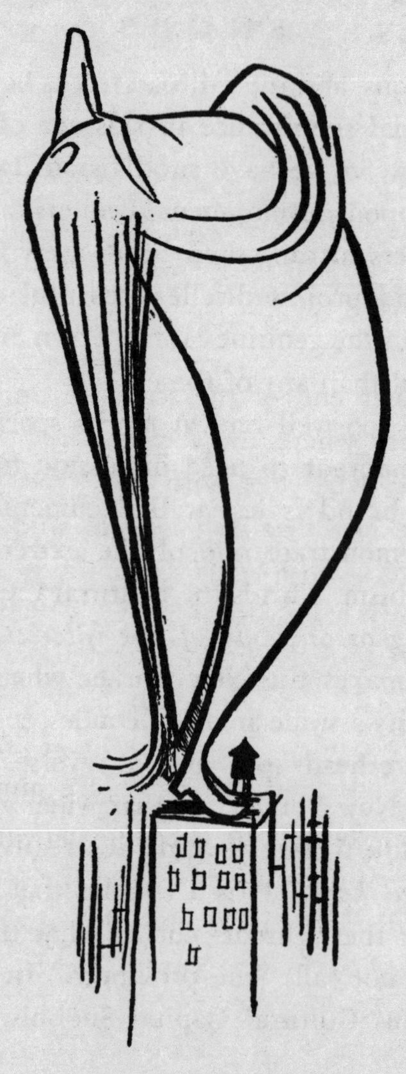

The Cultural Capital Snob will bend all the way over backward.

all the way over backward and touch its heels with its hair with some such observation as: "I think New Yorkers are the most provincial people in the world, don't you?" The born and bred New Yorker is rare (or at least thinks of himself as rare), and in general the New Yorker by adoption is the more virulent of the species.

At the other end of the scale we find the Small Town Variety where the *I have lived here longer than anyone* type vies with the type who makes much of the fact that only people who rub elbows with the members of a small community really understand the meaning of life. This latter type, like the Cultural Capital Snob, is usually a member of the community by adoption, having fled from the city in order to discover what he calls "real values." Sometimes the members of this group are summer people gone native who retain certain characteristic attributes of their type such as station wagons, and dress themselves in more elaborately rural costumes (blue jeans, checked wool shirts, even straw hats) than any genuinely rural inhabitant would consider proper or necessary. Another variant of this species is the ex-urbanite who buys a farm in order to "get next to the soil." These might be called the Eternal Verities

Snobs, Back to the Land Division, and are very likely to be authors.*

Before we proceed to our next category, there is one offshoot of the Regional Snobs which bears brief mention: the *world is my home* species† who pride themselves on the fact that they are as much at home in Shepheard's Hotel in Cairo as in the casino at Monte Carlo or in the Ritz Bar in Paris or in the Pump Room in Chicago or in less expensive saloons in any of these places. The members of this category like to think of themselves as "the international set" and are frequently remittance men, decayed nobility, career diplomats, overseas representatives (and their wives) of American industries, wealthy divorcees, or rich refugees. They regard every international problem or crisis merely as a personal inconvenience, and every visa in their passports as a mark of sophistication. The natives of any place they visit have no other function but to serve them, and their technique for insulting waiters is unsurpassed. Although the world is their home, they are in one sense the most provincial snobs of all, for their real world consists of a few

* Indeed, western Connecticut and Bucks County, Penna., have been so overrun by authors and editors that a real farmer can hardly afford to buy land there.

† Not to be confused with the One World Snobs (cf. Political Snobs, p. 50).

thousand wanderers, and their horizons are limited to the chips on the table, the bottles on the bar, the crystals in the chandeliers of hotel dining rooms, and when out-of-doors they darken their little world with sun glasses.

In a much diluted form this species is also found among those from urban areas who have been abroad at least twice but usually more often than that and whose conversation is peppered with such phrases as "When I was in Antibes last year . . ." or "Personally I prefer the Côte d'Azur," or "The food in Paris isn't what it was before the war." If they are culturally inclined, they are sooner or later likely to observe that "Nothing, but nothing, goes on in France. The exciting painters are all in Italy." They have an unmitigated scorn for all tourists, and are ashamed and embarrassed by their compatriots who travel abroad and try to avoid any identification or association with them.*

In general it can be said of the Regional Snob that while he may be tiresome he is usually innocuous. He is, however, practically unavoidable in one mani-

* There are two important variants of this species: (1) the Language Snob who pretends to five or six languages and sprinkles his conversation with French, Spanish, and German phrases, and (2) the Reverse Language Snob who prides himself on getting along everywhere with his native tongue on the assumption that anyone who doesn't know it is a fool or worse.

festation or another, and it is probable that as the world grows smaller, Regional Snobbism will increase. It is a logical antidote to political efforts to make man love his neighbor.

Moral Snobs

Like the Regional Snobs the number of Moral Snobs is legion and they love their neighbors no more dearly. Oscar Wilde, a really accomplished snob, said that "morality is simply the attitude we adopt towards people we personally dislike." But the Moral Snob carries it further than that; his snobbishness extends to people he doesn't even know. Morality is both a public and a private matter, to be sure, and it is characteristic of the Moral Snob to put a good deal of ornamental fretwork on his public façade and let the private places of his personality be slovenly. To call him a hypocrite would be to attribute vices to his virtues; he is not so positive a character as that. He does not necessarily want to get away with anything, but he is always quite sure that everyone else does, or would if he didn't keep a sharp eye on them.

In our day there are two main categories of Moral Snobs—the Religious Snobs and the Tolerance Snobs. In mentioning the former, I am aware that I am on delicate ground, but the Religious Snobs are identified with no particular sect or creed, and the true believer is rarely, if ever, snobbish about it. The only thing that all such Snobs seem to have in common is the conviction that those who disapprove of their faith or the methods by which they try to spread it are "bigots." Anyone who has been a member of a congregation is familiar with the church-goer whose attendance is prompted entirely by social or business considerations and whose snobbery is mainly directed at others similarly motivated but less meticulous about it. This type of Religious Snob (sometimes an *I pass the plate* Snob) has a special scorn for his friends who play golf on Sunday mornings. The female version of this type can be seen lingering after service in animated conversation with the clergyman or others of her own kind.*

Since the war we have seen a new manifestation of Religious Snobbery among intellectuals who in middle life have discovered religion. This group, the children of a generation of revolt against the Church,

* A different type of Religious Snobbery (non-existent in this country) is credited to the family of the Duc de Levis-Mirepoix, one of the oldest important French titles, that dates back to the ninth century. The family is purported to be descended from the sister of the Virgin Mary, and when the members of the Levis-Mirepoix family pray, they are said to say: "Ave Maria, ma cousine . . ."

may be humble enough when they are with other members of the congregation; their snobbishness manifests itself, however, when they are in the company of their friends who are agnostics. Their behavior is marked by what might be called the Private Discovery attitude toward God. Their conviction seems to come less from conversion than from a sense of invention, as though they had concocted the whole concept of belief, and they talk about God as though nobody had even heard of Him before, as though He were really an awfully interesting fellow that "you really should meet; you'd like him."

Sometimes opposed to the Religious Snob and sometimes allied with him is the Tolerance Snob, a species of comparatively recent origin. It should be noted that he turns the tables on the Religious Snob for lack of tolerance toward disbelievers and backsliders, and in such cases he often calls the Religious Snob a "bigot." The bigot is a most useful foil to the Tolerance Snob. But whether he is at loggerheads with the Religious Snob or not, the *I am more tolerant than anybody* Snob has a special predilection for getting his name printed on the letterheads of societies for the prevention and furthering of things, and not infrequently makes a career of the lecture platform and the composition of oh-the-injustice-of-

The Tolerance Snob has a special predilection for getting his name printed on letterheads.

it-all articles for magazines that specialize in tolerance. He is more likely to serve on committees than to take shifts at hospitals, and he is more prone to give money than time and labor to the causes he graces with his name.

There are, to be sure, many other familiar brands of Moral Snobs such as the *Life is earnest, night is coming* Snobs who look down with alarm upon those who dally with the pleasures of the flesh. They are likely to be troublemakers, as were the Prohibitionists, or nosey-Parkers like the Watch and Ward Society, or despoilers of frankness like Queen Victoria. There are the Pure Philosophy Snobs who sneer morally at those who dally with the frivolities of the mind, and who look upon popular culture as though it were a vice. There are also the Cleanliness Snobs, a category that includes everyone who believes that all foreigners are "dirty," as well as those who consider "daintiness" somehow identified with godliness.

It may be well to mention in passing the Anti-Moral and Amoral Snobs, a group not entirely confined to the organized and unorganized underworld. They are not uncommon in show business and on its periphery, and are found in a relatively pure form among carnival folk, boardwalk auctioneers, and inhabitants of Broadway bars. They devote their lives to "making a fast buck" and regard all who simply earn an honest living as "suckers" and "squares." These are the "sharpies," the "smart guys," to whom an honest penny is an ill-gotten gain and whose disdain for those who toil is unmitigated. Their pose is a combination of casual toughness and enigmatic superciliousness. Their philosophy is summed up in the phrase "get wise" or "wise up," and they are likely to indulge in "sharp" suits, yellow convertibles, and an extravagant show of devotion to "Mom."

The Moral Snobs are an uncomfortable lot, not only to themselves but to others, and potentially, as Gerald W. Johnson has pointed out in a recent article, they are a menace to society second only to that very rare species, the first-class political villain.

Sensual Snobs

In contrast with the Moral Snobs are the Sensual Snobs who take special pride in being able to wrest more pleasure per cell from the flesh than anyone else. In this general category, which is even more elastic than I mean to make it, we find the Food and Drink, the Sex, the Indolence, the Health and Hygiene, the Physical Prowess, and the Game Snobs. Presumably this category could be stretched to in-

The Sensual Snob takes special pride in being able to wrest more pleasure per cell from the flesh than anyone else.

clude snobs of all of the five senses, but the reader will quickly see that this would lead us into an endless dissertation. I will, therefore, merely suggest some of the more obvious specimens in hope that he may, by applying his own special insight, discover subtleties which I have missed and categories that he knows better than I.

The Food and Drink species is almost too common to require more than a passing word. In Food, the Herb Snobs, while somewhat old-fashioned, still persist, with their tidy rows of identical bottles from Wagners or Bellows or the House of Herbs containing not only exotic herbs but still more exotic peppers (Lampong, ground or unground, Nepal, etc.), smoked salt and other condiments. In the kitchen window is a pot of chives, and in the garden, if there is one, or even sometimes in a window box are varieties of fresh herbs—thyme, basil, and the like—for the real Herb Snob is at heart a *Fresh* Herb Snob. It is my observation that this species is less in the ascendancy now than the Pot Luck Snobs, Casserole Division, or the *This is something I just threw together at the last minute* species. The mussels-snails-brains-and-garlic group continues to operate, especially in areas where mussels, snails, brains, and garlic are still considered somewhat outrageous; and the

Plain American Food group ("if you want a good cup of coffee and a decent hamburger, eat in a diner") flourishes in metropolitan areas where good foreign cooking is a commonplace. The Foreign Food Snob often can be identified by his attitude of frustration. The "little place" that he has discovered and which used to be so good has always just recently gone to pot. "You know how it is," he says. "The frogs' legs Provençal used to be superb, but now the place has got popular, and the food isn't fit to eat any more."

The Drink Snobs are, of all categories, the easiest to identify since the rules are so well established. They insist that their whiskey be bonded; they know what proof it is, and they drink it neat or "on rocks"; their Scotch is "V.O." or "V.V.O."; their martinis are as dry as almost no vermouth can make them (in restaurants where they suspect the martinis may be somewhat amber in hue they order Gibsons and remove the onions), and they always nod at the waiter after looking at the date on a bottle of wine. Only genuine connoisseurs have the self-possession to send back a bottle of wine. Some Drink Snobs take special pride in the amount they can consume and not show it; others take special pride in having a worse hang-over than anybody ever had before.

The Sex Snobs have been adequately documented by the Physiology Department of Indiana University. It may, however, be interesting to note that the publication of Dr. Kinsey's first volume on *The Sex Habits of the Human Male* produced two new manifestations of the Sex Snob; first, those of the *I could tell Kinsey a thing or two* variety; and, second, the species that insisted that the excitement about the book was all nonsense—"Why, I've known all that for years." The attitude of the British toward the Kinsey Report reveals an interesting provincialism. I was told by Dr. Kinsey that in general the reaction of the British professional and, if I may be permitted the phrase, the lay press, was: "No doubt this is all very true about Americans, but we are not interested. The British don't behave like that."

The Indolence Snobs, on the other hand, have been epitomized by an Englishman, Cyril Connolly, in his book *The Unquiet Grave.* "Others merely live," he wrote, "I vegetate." An interesting counterpart to Mr. Connolly's form of snobbism is to be found in those who make a great show of doing nothing, of sleeping late, of lying in the sun, of always having time to amuse themselves and their friends, and who at the same time produce a great deal of work. These are the people who express their superiority by saying, "I just tossed off this novel in my spare time," or "I just thought of this new international trade combine over a game of canasta in Miami one evening."

The Health and Hygiene Snobs may not properly belong with the Sensual Snobs. It is possible that they belong more comfortably as a subdivision of the Moral Snobs, or possibly they fit better with the Physical Prowess Snobs whom we shall discuss in a moment. There is no denying, however, that there is sensual pleasure in the subjugation of the flesh, and that this is part of the routine behavior of the Health and Hygiene Snob. It is a far stronger motive than mere laziness that keeps a man or woman flat on his face in the hot sun for a few hours in order to turn first red and then brown; it is certainly not morality that sends men and women to gymnasiums to reduce one portion of the anatomy and exaggerate another, nor is it laziness that makes them diet, abstain from (or at least be ostentatiously moderate about) liquor, and get to bed at what they call "a reasonable hour." It is the delight of being able to look down upon those who, to use their phrase, "don't take proper care of themselves." Sex, of course, enters strongly into this, but then so does a feeling of moral superiority. I have no doubt that the social scientists will in time be able to isolate the Health and Hygiene Snob from the Moral Snob.

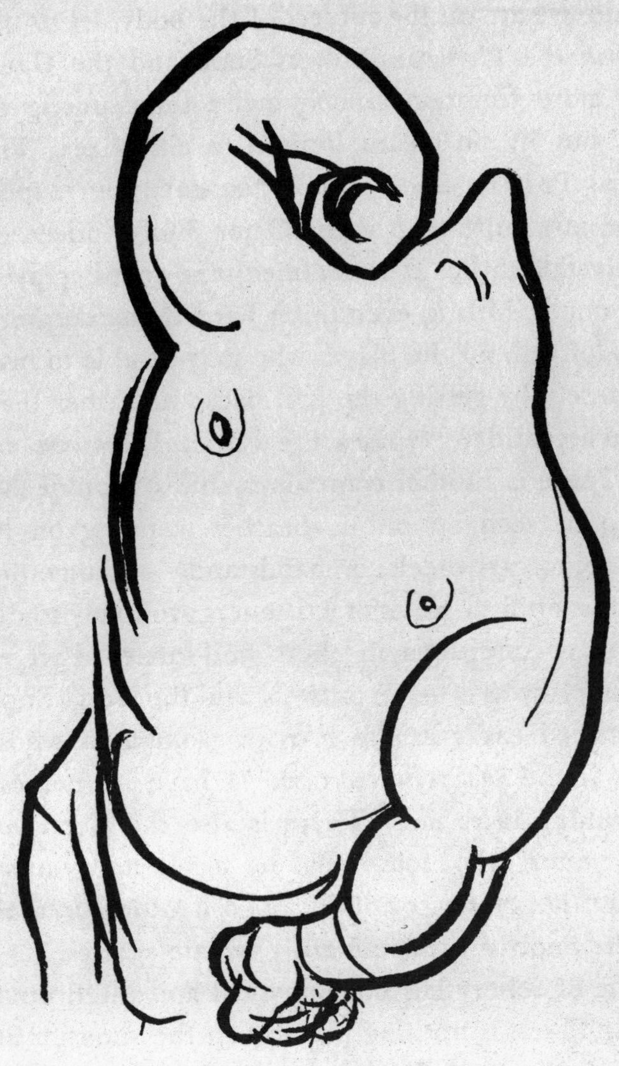

The Physical Prowess Snob is likely to die of a cardiac condition in his mid-forties.

While we are on the subject of the body, let us not overlook the Physical Prowess Snob and the Game Snob, more common among males than among females, but by no means limited to either sex. The Physical Prowess Snob is not necessarily an expert athlete, any more than is the Game Snob; indeed he is likely not to be. It is the mediocre tennis player, for example, hitting everything hard if inaccurately, who is lofty about the player who may be able to beat him merely by getting the ball back; he rather than the varsity athlete typifies the Physical Prowess variety. There is another common manifestation of this type to be seen on public beaches standing on its head, doing cartwheels, or handstands—a dangerous form of snobbism, as its practitioners are likely to die of cardiac conditions in their mid-forties. Part of physical prowess is nerve prowess, and the Nerve Snob is identified easily as the man or woman in whose mouth is the recurring phrase: "I have no nerves" or "Nothing fazes me." There is also the *Strong as a horse* Snob who boasts that he never needs more than four hours' sleep, and his opposite, the *Tired all the time* Snob, a predominantly female species.

Game Snobbery has both physical and intellectual aspects. There is nothing physical in the superior attitude of the chess player to the bridge player, to be sure, unless it can be said that the emotions are of physical origin and that scorn is an emotion. The same may be said of the attitude of the golfer to the bowler, of the billiards player to the pool player, and of the polo player to all other players of any sort whatsoever. The predominantly sensual aspects of Game Snobbery come into play only in the dangerous sports—football, lacrosse, bullfighting, professional hockey, and other socially acceptable means of blood-letting.*

There is a distinction to be drawn between gamesters and sportsmen, between athletes and out-of-doors men and women. In contrast to the gamesters who compete with one another, these sensualists compete only with nature—with mountains and seas and prairies—and with the unsuspecting doe and buck, the pheasant, the trout and the grouse. But there are two quite distinct types of Sportsmen Snobs—the Out-of-Doors Snob and the Great Out-of-Doors Snob. The former is essentially a suburban backyard or barbecue type who limits his outdoor activities to picnics and outings in the family car. He will tell you how much better a steak cooked over his grill in the

* The foremost proponent of this type of snobbery, combining its blood-letting, literary, and chest-thumping aspects is surely Ernest Hemingway.

backyard tastes than it does if it is cooked forty feet away in the kitchen. While converting the steak into charcoal, he sometimes wears an apron with slogans on it and a chef's cap.

The Great Out-of-Doors Snob, on the other hand, talks of himself as a "man's man" because he insists that men can only be men in the presence of their own kind and a long way from home. To him there is nothing more cleansing to the spirit than to sleep on balsam boughs in "the great green cathedral of the woods," nothing builds character like the precipitous slopes of a mountain, and nothing so entrances the palate as a trout he has caught on a dry fly. To him all country that is four hours drive from home is "God's country." It is not surprising that the most snobbish Great Out-of-Doors Snobs earn their livings and spend the greatest part of their lives in cities as far removed from nature as possible.

No matter what you may think of the Sensual Snobs, it cannot be denied that, unlike the Moral Snobs, they are a great pleasure to themselves.

Emotional Snobs

Since the emotions carry us rapidly in dangerous directions and soon lead us to the darkest corners of

The Emotional Snobs, or I feel things more deeply than anybody *variety.*

man's nature, we must proceed to the dissection of the Emotional Snobs with caution. This is the *I feel things more deeply than anybody* variety, and there is likely to be at least one in every family.

Probably the largest single subdivision of this category is the Love Snob, a type which finds its roots among adolescents, who, since they are meeting their first encounter with sexual love, believe that no one has ever been so in love before. Their intolerance of their juniors is matched only by their scorn for their elders, and this can set a pattern for adult love that is difficult to break. The so-called "great lovers" do not, I believe, belong in the Love Snob category but rather in that of the Sex Snob. It was surely not about the intensity of his emotions that Don Giovanni, with his list of eleven hundred ladies, was vain.

The Mother Love Snob, or *I give my all for my children* type, is not uncommon among women who are not Sex Snobs, and it is probable that the second volume of the Kinsey Report may shed some light on this. The Filial Love Snob, or Mom Snob, is not in my experience nearly as common as English authors, such as Geoffrey Gorer, or Americans such as Philip Wylie contend that it is. That is not to say that the exploitation of Mom Snobbery by the florists once a year does not give it at least a seasonal boost. The

Sibling Love Snob probably does not exist, and the Brotherly Love Snob belongs more properly under the Moral Snobs than under Emotional Snobs.

The Marital and/or Soul Mate Snobs are not rare, though they are particularly tiresome because they are, by the very nature of their snobbery, raised to a higher power. Since it takes two to make soul mates, they are twice as tiresome as other snobs.

The Popularity Snobs also belong in the Emotional group; in a sense they are everybody's soul mate. To use their own vernacular, they have "a way with people" and can "get along with anybody." Theirs is the hauteur of affable condescension, and traditionally the species is common among traveling salesmen, Rotarians, public relations counselors, and politicians, though it would be a mistake not to recognize the far wider ramifications of this type wherever we meet them. Mass demonstrations of Popularity Snobbism are known as conventions.* The typical member of this species rarely uses the form "mister" in addressing anyone, no matter how brief or perfunctory the acquaintance. He is strictly a first name man, and has little respect for anyone's dignity or

* College reunions also figure in this category. They provide opportunities for the temporary renewal of Popularity Snobbism in those who were popular in college but have been slipping ever since.

privacy. He assumes that everybody loves him, and he reasons that there is no privacy in a public love affair.

By contrast the Unpopularity Snob, or *Nobody can get along with me* type, takes two principal forms.* The first is an imperious and often petulant species who by dint of the loftiness of his position or intellect makes much of the fact that he can't be bothered with boors and idiots. He works with his door closed; he throws all second-class mail into the wastebasket without opening it, and he never seems to be able to remember anyone's name or if he does he mispronounces it. When you meet him, he says "Hello," but looks past you, as though you were obstructing his view. The second is the sensitive or *I'm too special* type who is "misunderstood" by crass and materialistic people. This species is likely to gravitate in the general direction of the arts and crafts and sooner or later to metropolitan areas.

Before leaving the Emotional Snobs let us merely list a few other common varieties: the Pain Snobs,

* The persistence of one type of Unpopularity Snob is demonstrated by the number of adults who take special care to make the point that anybody who amounts to anything was "unhappy in school." It is likely to express itself in some such direct statement as "I was the most unpopular boy (girl) in my class."

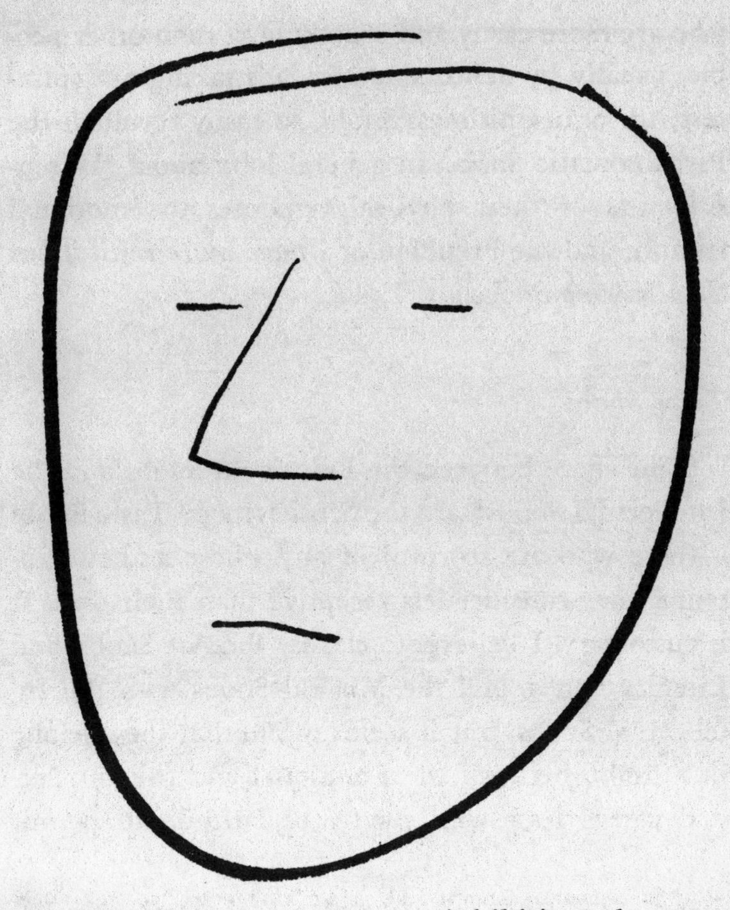

The Freudian, or I have more inhibitions than anyone *Snob*.

who are more easily and acutely hurt than other people, usually by dentists and friends taking out splinters; the Squeamishness Snobs, so easily revolted; the Psychosomatic Snobs, in general lofty about the sensitiveness of their physical responses to emotional stimuli; and the Freudian or *I have more inhibitions than anyone* Snobs.

Taste Snobs

Somewhere between the Emotional Snobs and the Intellectual Snobs* are the Sensitivity or Taste Snobs—those who are scornful of any whose æsthetic antennæ they consider less receptive than their own. It is customary, I believe, to classify the Art Snobs, and Literary Snobs, and the Musical Snobs with the Intellectual Snobs, but it seems to me that they belong in a limbo between the Emotional and the Intellectual categories,† with plenty of latitude to permit

* So commonly known and, as we have noted, so well established as to need no discussion in this brief survey.

† In his book *Varieties of Delinquent Youth,* Dr. William H. Sheldon approaches one fairly typical species of Taste Snob from a different direction. He refers to them as DAMP RATS, a name devised from an acrostic of their characteristics, as follows: D—Dilettante, A—Arty, M—Monotophobic, P—Perverse; R—Restive, A—Affected, T—Theatrical.

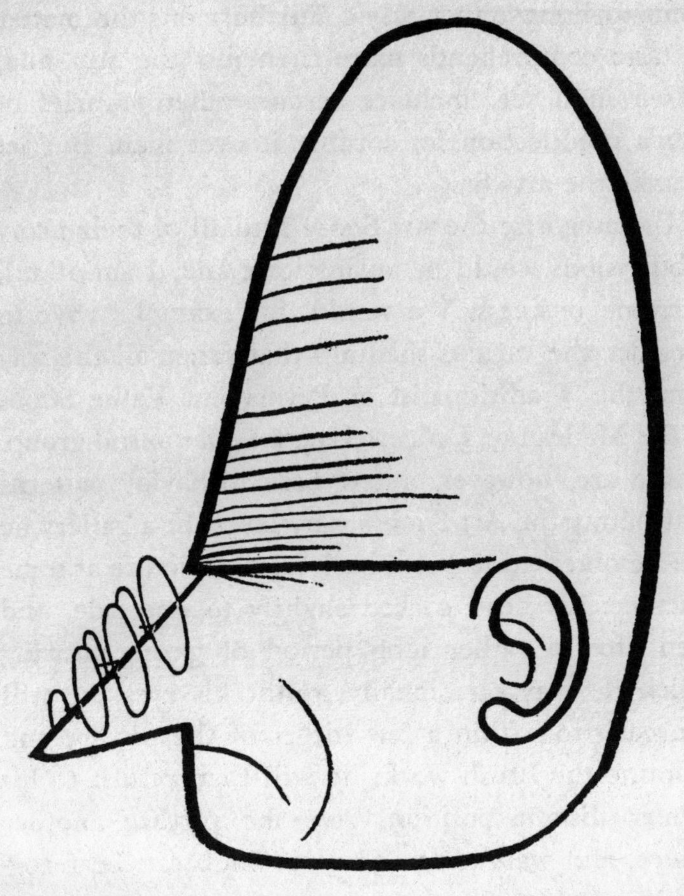

The Intellectual Snob is so well established as to need no discussion in this brief survey.

them to jump either way.* Furthermore the matter of taste comprehends more than just the arts, and, as we shall see, includes certain other vagaries of man's predilection for lording it over man. But let us take the arts first.

To categorize the Art Snobs into all of their many subdivisions would be an intricate and, I am afraid, tiresome business. We would, for example, have to consider the various shadings that range all the way from the Traditionalist or Permanent Value Snobs to the Modern or *I always keep an open mind* group. There are, however, a few basic behavior patterns that betray the Art Snob at any level. In a gallery he can be observed to stand back from a picture at some distance, his head cocked slightly to one side, and then after a rather long period of gazing (during which he may occasionally squint his eyes) he will approach to within a few inches of the picture and examine the brush work; he will then return to his former distant position, give the picture another glance, and walk away. The Art Snob can be recog-

* It is interesting to note in connection with the publication of a new typographically eccentric magazine called *Flair*, that as an undergraduate Thackeray at Cambridge in 1829 contributed to a magazine called *The Snob*. An advertisement described it as follows: "Each number contained only six pages . . . printed on tinted paper of different colors, green, pink, and yellow." *The Snob* lasted for only eleven numbers.

nized in the home (*i.e.*, your home) by the quick look he gives the pictures on your walls, quick but penetrating, as though he were undressing them. This is followed either by complete and obviously pained silence or by a comment such as, "That's really a very pleasant little water color you have there." In his own house his manner is also slightly deprecating. If you admire a print on his wall, he is likely to say, "I'm glad you like it. It's really not bad considering it is such a late impression." Or if he is in the uppermost reaches of Art Snobs and owns an "old master" which you admire, he will say, "Of course Berenson lists it as a Barna da Siena, but I've never satisfied myself that it isn't from the hand of one of his pupils."

The Literary Snob has not only read the book you are reading but takes pleasure in telling you the names of all the earlier and more obscure books by the same author, and why each one was superior to the better known one that has come to your attention. His attitude toward best sellers may take two forms: either he reads best sellers because he "likes to know what eye-wash the public is being fed these days," or he doesn't read them at all because he knows that "obviously no book worth bothering with is going to appeal to that many semi-illiterates," or "if it's worth bothering with it will outlast its vogue, and I'll read it later." He wouldn't think of reading the short version of Toynbee, for example, any more than he would the *Reader's Digest*; he considers this to be spoon-feeding and beneath his intellectual dignity. He doesn't bother with the "latest thing" with one notable exception; he makes a great point of "the latest thing" imported from English publishers: the "new" Elizabeth Bowen, the "new" Edith Sitwell, and the "new" Ivy Compton-Burnett. He may keep *au courant* with the literary columns of the *New Statesmen*, and he makes moan over the recent passing of the English literary magazine, *Horizon*. A lower order of Literary Snob, the Thursday Morning Club or Reading Circle type, has always read the current best-sellers, and his (or more likely her) scorn for anyone who has not read them easily matches the scorn of those we have just mentioned who would not read them.

Musical Snobs are in general of three sorts—Classical Snobs, Jazz Snobs and Folk Snobs. The first can sometimes be identified at concerts because they keep their eyes closed. This can for obvious reasons be misleading, but if closed eyes are accompanied by a regular movement of the hands in time with the music, it is clear that the listener is beating time to himself. This is characteristic of the lower orders of

Classical Snob. If he has a score of the music which he follows while it is being played, he may be a professional musician looking for subtleties of interpretation; he may, on the other hand, merely be a higher order of Classical Snob. The surest way to identify the Classical Snob is to see whether he comes back after the intermission or not; if he stays only for the more difficult or abstruse part of the program and ignores the more popular portion, he is either a snob or a professional critic, or possibly both.

Musical Snobs, Jazz Division, beat time not with their hands but with their feet. They do not talk about records or recordings but about specific choruses, solo passages, or "breaks." They know the dates and numbers of original pressings and occasionally they collect never-played records much the way some book collectors prefer rare copies with uncut pages. They are well grounded in the brand of jazz they refer to as "authentic" (New Orleans, Memphis, Chicago) and they are extremely partisan about what they consider to be "advanced" (Progressive Jazz, Bebop, or even Dixieland). There are some overtones of Social and Racial Snobbery in the way Jazz Snobs identify themselves with jazz musicians.

There is, of course, an affinity between the Jazz Snobs and the Folk Music Snobs, since both types are committed to the area of popular culture; but the affinity seems to stop there. The Folk Song Snob regards the deep cooing of Burl Ives as mere popularization and vulgarization of essentially authentic ballads; he prefers the falsetto of John Jacob Niles (who makes his own dulcimers and sports a bandana), but gives his heart only to the thin whine of the genuine mountaineers, whose songs have been recorded by Alan Lomax for the Library of Congress in their native habitat. Cowboy songs in general are considered less arresting than mountain songs, and such throwbacks as "Mule Train" are regarded as aggressively new and hence beneath contempt. Only the lower orders of Folk Snob actually engage in square dancing; the real snob sits by and evaluates the technique of the "caller."

To some Taste Snobs the movies are also known as a Folk Art, but to the real Movie Snob they have the *cachet* of the Fine Arts. They refer to them as either "the films" or "the cinema," preferably the latter, and they have little use for any that are not in a foreign language with English titles. Their contempt for Hollywood is the glue that binds them together into a strong little band of "serious" appreciators. Many of them belong to "film societies" and subscribe to series of "screenings." Second to

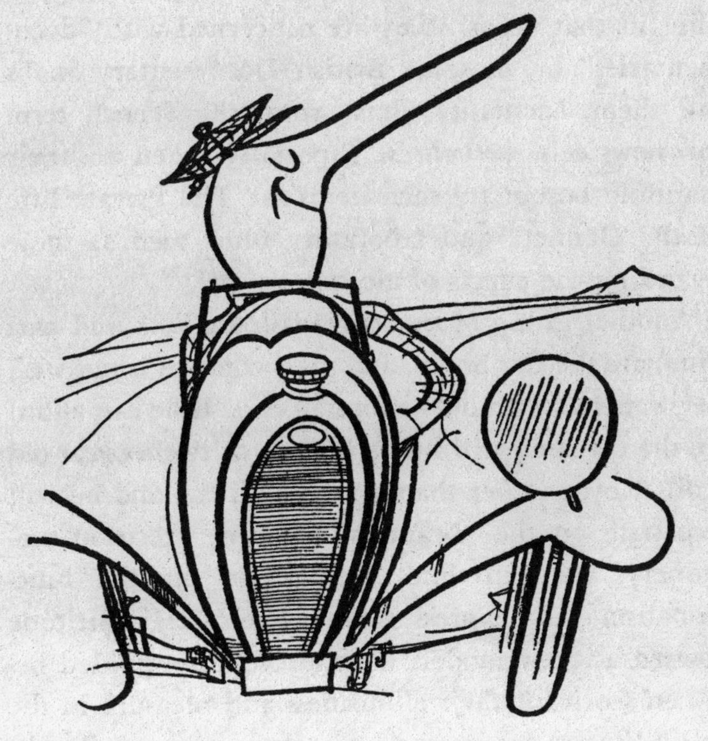

The Car Snob's concern is with old or foreign cars.

Italian, French, German, and an occasional English film (in that order), they are concerned with "documentaries" or, as some British Documentary Snobs call them, "actuality films" from the French term for newsreels, *actualités*. Especially taken to their manifold bosom are such items as "The Private Life of the Gannet" and laboratory films such as those recording the effects of alcohol on cats.

Another group of æstheticians (part Folk and part Fine) are the Car Snobs, and their concern is not with new cars but with old or foreign cars. If he can afford to, the car snob will buy a fifteen- or twenty-year-old Rolls Royce rather than a new Cadillac, and he will expatiate on the "beautiful piece of precision machinery," the "pure lines of the design" and the "functionalism that expresses the wheels." His attitude toward all new models is about equally divided between scorn for their clumsiness and pleasure in the fallibility of those who are taken in by "all that hideous chromium bright-work." He may be an Old Car Snob, or *Nobody has ever built a car as good as the Model-T* type, but in any case he looks down upon those who either just want a car to get them around or those who buy in order to keep up with the neighbors. If he were to buy a new car himself, it would be a British M-G or possibly a Cis-Italia.

Also among the Taste Snobs are to be found the Clothes Snobs, both male and female. In this instance the female is a good deal more interesting and varied than the male, for while the male "sharp dressers" are snobs of a sort, there is only one male Dress Snob who needs to arrest our attention: the Conservative Dress Snob. The buttons on the sleeves of his jacket actually unbutton. There is no padding on his shoulders. The collar of his shirt is a little too high for him, so that it bulges and wrinkles slightly, and it buttons down. He cares deeply about good leather and good tweed, but most of all he cares about being conspicuously inconspicuous.

The female Dress Snobs offer a far more complicated range of types, and it requires some temerity on the part of a man to broach this subject at all. In general, however, women seem to fall into the following categories of sartorial superiority:

a. The Underdressed Snob, who wouldn't be caught dead at a cocktail party in a cocktail dress, and a similar type, the next on our list . . .

b. The Basic Dress Snob, who believes that she has so much personality that she can get away anywhere in a simple black ("basic") dress and one piece of "heirloom" jewelry.

c. The Good Quality Snob, or wearer of muted

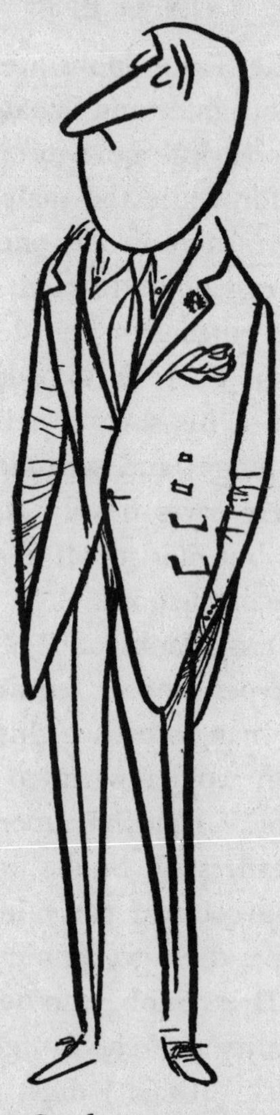

The Male Clothes Snob cares most of all about being conspicuously inconspicuous.

tweeds, cut almost exactly the same from year to year, often with a hat of the same material. This type is native to the Boston North Shore, the Chicago North Shore, the North Shore of Long Island, to Westchester County, the Philadelphia Main Line, and the Peninsula area of San Francisco, etc. It rides horses and is rare in Southern California, except for Pasadena. In Texas it trades at Neiman-Marcus.

d. The Band Box Snob—common among professional fashion models and among other young women trying to make their way in the big city. They look as though they had just stepped out of *Vogue* or *Mademoiselle*. They are never ahead of the fashion, but they are screamingly up to date.

e. The Dowdy Snobs, or *Who the hell cares about fashion* Snobs.

f. The Personal Style Snobs, or *I know more about my type than the experts* Snob. This final type considers her taste to be above the whims of mere fashion. She is so chic that she believes that it is unchic to be merely fashionable.*

The Taste Snobs form a group, unlike many of the others, which is a combination of both professionals and amateurs. The professional taste-dispensers—ar-

* The outstanding example of this in our time is the Queen Mother of England.

chitects, interior decorators, curators, dress designers, editors, advertising executives—are just as likely to be found in this category as are consumers.

Good taste is everyone's prerogative (no one willingly confesses to bad taste), and so nearly everyone is a Taste Snob of one sort or another, and often of many sorts at the same time.

Occupational Snobs

Our next category, the Occupational or Job Snobs, are of two sorts—those who are snobbish about the kind of occupation by which they live, and those who are snobbish about how they perform in their occupation. Few women, for example, are snobbish about being housekeepers; many are snobbish about the way they keep house. Many men, on the other hand, are snobbish about the positions they hold and less snobbish about how they perform in them. But first let's take the women. The woman whose dearest ambition is an absolutely well-ordered and efficiently run house looks down upon the woman who firmly believes that it is nonsense to spend so much time over the household that there is no time for what she calls "life." She in turn looks down upon the whole-souled housekeeper. It boils down to a conflict between two aphorisms—"cleanliness is next to godliness" and "a little dirt never hurt anybody," which, if we weren't careful, would lead us back to our discussion of Moral Snobs. Of course both of these types are looked upon with scorn by the Female Career Snob who manages with overbearing aplomb both a job and a household.

The Female Career Snob, however, is a novice compared with the Homemaker type, or *I just make a cozy place for my family to come home to* species. This variety knows no geographical limitations, in fact very few limitations of any sort. Especially it knows no limitations on the number of children it bears. The Homemaker Snob looks down upon all women who do not have four children or more and considers that no woman is doing her duty to society unless her home somewhat resembles a one room school at recess—children everywhere, their whittling, paper-cutting, water color smudges cluttering the mantel, the stairs, the piano. "I just pick up after them," this type says. "We rub along together—just one big happy family."

The hierarchy within which men work is quite different, and makes quite different demands. The professional man feels somewhat lordly toward the businessman or "money grubber" and considers him lacking in sensibility, intellectual curiosity, and near-sighted to the point of seeing nothing beyond the sales chart but the golf course or the bridge table. He is likely to blame the world's ills on the businessman's greed and lack of cultural understanding. The businessman, on the other hand, thinks of many professional men as "dreamers" and "idealists" or even as "pantie-waists." This applies especially to artists, writers, actors, musicians, scholars, and editors. The businessman is less likely to be snobbish about physicians, lawyers, and engineers because he considers them, like himself, to be "practical men." His most unlimited scorn is for bureaucrats who "have never met a payroll."

Performance on the job is less likely to matter than position, as I have said, but there is the Efficiency Snob whose pose is primarily one of crispness. He answers the phone by barking just his last name. He is inclined to have rows of buttons on his telephones or desk, and almost no papers. His memoranda are brief to the point of being curt, and he considers the word "please" something that has no place among desks and typewriters and he wants things done "soonest." He thinks of himself as a "trouble-shooter" and makes lists of possible troubles to shoot. As each one is shot, it is crossed off the list with a firm black line. Accomplishment is measured by the number of black lines, and everyone who doesn't measure up to his particular standards of efficiency is "hopeless." The reverse of this type, also common, is the man who lives behind a mess of papers, pencils, paper clips and "can never find anything" and yet manages to get out the work. The results produced by the Efficiency Snob and the Inefficiency Snob are much the same.

Before leaving the Occupational Snob, there are several allied types that bear mention. One is the Horse's Mouth Snob who always has a "pipe line" to the top for any information he may impart, and he is closely related to the type who always knows "just the man" to do any sort of job.

Performance off the job often reveals the Manual Dexterity Snob who can do complicated mechanical things with his hands and who considers all who can't to be fumbling idiots, and the opposite of this, the All Thumbs Snob, commonly found among women. Men who are all thumbs are sometimes reticent about it; women rarely are.

Political Snobs

Since political views and occupations seem to have a certain if not a specific affinity, let us next consider our penultimate category, the Political Snobs. It is well to note, however, that Political and Social Snobbery sometimes become confused. In the days when President Roosevelt was generally known in some Republican circles as a "traitor to his class" the overtones of Social Snobbery were strong, and the same applies, though in a diametrically opposite way, in the case of Henry Wallace's slogan, "The Century of the Common Man." This is rather more an example of Reverse Social Snobbery than of Political Snobbery.

Genuine Political Snobbery breaks down into five main categories: the Realistic Snobs, the Idealistic Snobs, the One World Snobs,* the Political Know-How Snobs, and the Anti-Political or *It's all a dirty business* Snobs. Each of these types, except the last, is to be found in every political party from the extreme left to the extreme right. The Realistic Snobs, for example, are extremely impatient with the Idealistic Snobs whom they refer to as "fuzzy-minded" or "muddle-headed" while they themselves put their faith in "practical politics." The Idealistic Snob is often brushed off by the left as "unreliable," by the right as "liberal," and by the middle-of-the-roaders as "impractical," while he in turn scorns the others as "short-sighted" or "sold out to the interests" or to

* Not to be confused with The World is My Home Snobs (cf. Regional Snobs p. 12).

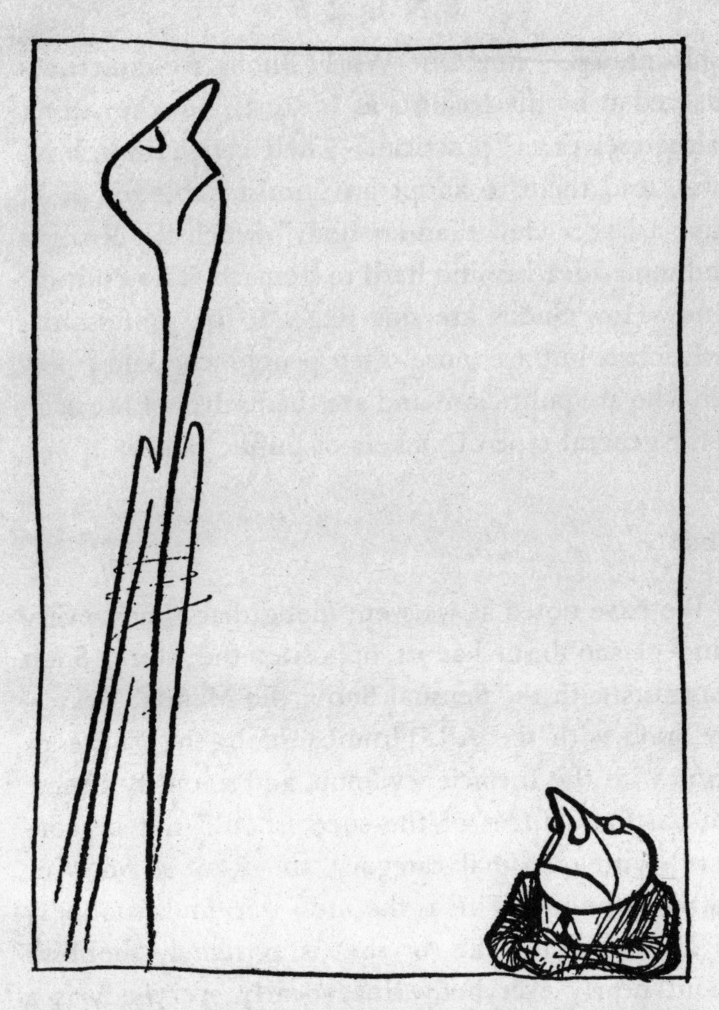

The Realistic Political Snob is extremely impatient with the Idealistic Political Snob and vice versa.

"self-interest." The One World Snobs are sometimes sneered at by the Realists as Idealists, but they think of themselves as "practical." Their convictions, however, lead them to adopt an unmistakable air of "I have a larger view than anybody" which the Realists and some Idealists find hard to stomach. The Political Know-How Snobs are not likely to be professional politicians but are more often people who know people who are politicians and are themselves of the armchair general type of makers of public policy.

Coda

We have noted as we went along that almost every kind of snobbism has its opposite: the Moral Snob contrasts with the Sensual Snob; the Manual Dexterity Snob with the All Thumbs Snob; the Efficiency Snob with the Inefficiency Snob, and so on. But these contrasting sources of the sneer should not be confused with our final category, the Reverse Snob or Anti-Snob Snob. This is the snob who finds snobbery so distasteful that he (or she) is extremely snobbish about nearly everybody since nearly everybody is a snob about something. This is the man who tries so hard to be "natural," so hard to be "just folks," so hard to avoid having anyone else think he is a snob, that he plays a game which (if I may be forgiven for being a Language Snob for a moment) is *faux naif.* He would not, for example, ever be caught using a foreign phrase, as I have, lest it be thought pretentious even when it serves better than any other he can think of to convey his meaning. Or if he is forced to use it (or even a foreign name, let's say) he Americanizes its pronunciation lest anyone think him upstage.* He makes much of the fact that simple, uneducated people are wiser and nicer than sophisticated and educated people, even wise and nice educated people. He plays down his own education and accomplishments with an elaborate display of modesty and is likely to introduce a very erudite and perceptive observation with the phrase "Of course I know so little about this I have no right to an opinion," or "I know this is probably stupid of me, but . . ." Of all the snobs the Reverse Snob is possibly the most snobbish; he is so sure of himself that he intentionally puts other people in a position where they have to play his game or feel like snobs themselves. The false simplicity of the Reverse Snob stands in direct and glaring contrast to the genuine simplicity of the genuinely modest man.

By and large it is only the very great who are not

* "They spell it Vinci and pronounce it Vinchy; foreigners always spell better than they pronounce."—Mark Twain, *The Innocents Abroad.*

snobbish at all. They are the ones who are modest about their accomplishments because they have devoted their lives to achieving some kind of understanding and so have developed a deep tolerance for ignorance. By the same token the serious professionals in any field are not likely to be snobbish about other serious professionals, whether they are doctors or actors or writers or mechanics or businessmen or masons or even, let it be said, housekeepers. As we noted at the outset, it is those who are unsure of themselves and are seeking security in their social relationships who have provided us with this incomplete list of Snobs.

It will not have escaped the reader (and so I might as well admit it) that this cursory attempt to classify and define snobs is an example not only of Intellectual Snobbism but of Moral, Sensual, Occupational, Political, Emotional, and above all of Reverse or Anti-Snob Snobbism. I am sure there is no greater snob than a snob who thinks he can define a snob.